Problem Framing for Measurable Impact

(PFFMI)

Problem Framing for Measurable Impact

(PFFMI)

Transforming IT Documentation

D Horton

Recentria Press

An Imprint of Abolitic

Title: *Problem Framing for Measurable Impact* (PFFMI)
Author: D Horton
Publisher: Recentria Press, an imprint of Abolitic
Book design by: Daryl Horton
Printed in the United States
First Edition

ISBNs
Paperback: 978-1-945047-30-5
Kindle: 978-1-945047-31-2

This book is a work of nonfiction. The sermons and reflections presented are the author's original words and interpretations, offered for spiritual insight, historical context, and the encouragement of thoughtful reflection. Military titles and ranks are used for historical and professional accuracy and do not imply endorsement by the Department of Defense.

For inquiries, permissions, or bulk orders, contact the publisher at: www.abolitic.com/

CONTENTS

INTRODUCTION

Charting New Territory

Accomplishment Statements as Organizational Knowledge

For too long, IT Service Management has treated accomplishment documentation as a necessary administrative task – a record of what was done. While valuable for accountability, this approach often misses the critical opportunity to transform routine reports into powerful engines of organizational learning and innovation.

This book represents the *first foray* into a fundamentally new way of thinking about accomplishments. We're moving beyond simple reflection and towards a dynamic system where accomplishment statements aren't just documented – they are actively *created, managed, and leveraged as key organizational knowledge products.*

As we've outlined in the CI-Driven PFFMI framework, this approach recognizes that actual value lies not simply in achieving outcomes, but in understanding *why* those outcomes were achieved, what lessons can be learned, and how that knowledge can be shared across the organization.

This isn't a quick fix; it's a complex undertaking. It demands a shift in mindset – from viewing accomplishments as isolated events to recognizing them as interconnected nodes within a broader service ecosystem. It requires integrating robust Knowledge Management capabilities, fostering collaborative learning loops, and empowering teams to share their expertise proactively.

This book will provide the foundational principles and practical guidance for embarking on this journey. We intend to lay the groundwork for a more agile, resilient, and innovative IT Service Management practice – one where every accomplishment contributes not only to operational efficiency but also to sustained organizational growth.

We acknowledge that this is an initial exploration. The full potential of this approach will be realized through ongoing experimentation, adaptation, and collaboration across your organization. Consider this a starting point – a foundation upon which you can build a truly self-organizing system of knowledge and learning.

ONE
The Problem with Traditional Reporting

Let's be honest: how many times have you read an accomplishments report and felt... vaguely unsatisfied? You see metrics, numbers, and descriptions of tasks completed, but the *real* value—the demonstrable impact on the business—often feels obscured. In IT Service Management (ITSM), documenting achievements isn't just a procedural requirement; it's a critical tool for justifying investments, demonstrating efficiency, and driving continuous improvement. However, too often, accomplishments reporting falls short of this potential, leading to frustration, wasted effort, and ultimately, a disconnect between what teams are doing and the value they deliver.

The Symptoms of a Broken System

We've all encountered it: the report that lists "Resolved 123 Tickets" without context. It doesn't tell us *which* tickets, *why* they were resolved, or *what impact* those resolutions had on our customers or business operations. Here are some common symptoms of a broken accomplishments reporting system:

- Lack of Context: Reports often lack sufficient background information, making it difficult to understand the significance of the achievements.
- Focus on Outputs, Not Outcomes: The emphasis is frequently placed on *what* was done (e.g., "Implemented new monitoring tool") rather than *why* it was done and *how* it benefited the business (e.g., "Reduced downtime by 15%").
- Missing Traceability: It's often impossible to trace a specific achievement back to its root cause or the team responsible. This hinders learning and prevents effective problem-solving.
- Generic Statements: Reports are filled with vague, buzzword-laden statements that lack concrete detail ("Improved service levels," "Enhanced user satisfaction") without quantifiable evidence.
- Data Silos: Information is often scattered across multiple systems, making it difficult to get a holistic view of performance.

Real-World Examples

Let's look at some scenarios:

- The "We Fixed the Printer" Report: A support team resolves printer issues – great! But without understanding *why* the printers were malfunctioning, *how many* users were affected, or *what the impact was on productivity*, it's just a number.
- The "New Software Deployment" Report: A new software is deployed – fantastic! However, if the report doesn't detail the user training provided, the post-deployment support required, or the resulting improvements in efficiency, the value remains unclear.

The "Incident Resolution" Report (Without Root Cause Analysis): Simply stating "X number of incidents were resolved" without identifying the underlying causes and implementing preventative measures is a missed opportunity for genuine improvement.

Conclusion

These examples highlight a fundamental problem: traditional accomplishments reporting often fails to capture the *actual* value of IT service management efforts. It's time for a new approach – one that focuses on systems thinking, measurable impact, and traceability. This book will introduce you to CI-Driven PFFMI, a framework designed to transform your accomplishments documentation from a tedious chore into a powerful tool for driving strategic success.

TWO
Systems Thinking – A Foundational Concept

Before diving into the specifics of CI-Driven PFFMI, it's crucial to understand its foundation: Systems Thinking. It's not just a buzzword; it's a fundamentally different way of looking at problems and solutions. Instead of focusing on isolated components, systems thinking emphasizes understanding how things are interconnected – how actions in one area ripple through the entire system and affect others. It's about recognizing that complex issues rarely have simple, linear causes and effects. Mastering this concept will be key to unlocking the full potential of PFFMI.

What is Systems Thinking?

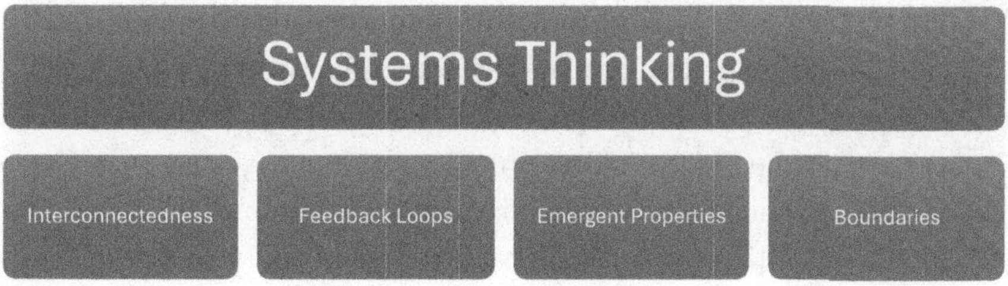

At its core, systems thinking is a holistic approach to problem-solving. Here's a breakdown of the key elements:

- Interconnectedness: Everything within a system is linked – resources, processes, people, and technology. Changes in one area will inevitably impact others.
- Feedback Loops: Systems operate through feedback loops – where outputs become inputs, creating continuous cycles of influence. These can be positive (amplifying change) or negative (stabilizing the system).
- Emergent Properties: Complex systems exhibit properties that aren't present in their individual components. The whole is *more* than the sum of its parts. For example, a team working together can achieve far more than the same individuals working independently.
- Boundaries: Systems have boundaries – defining what's inside and outside the system being considered. It's important to identify these boundaries to avoid getting lost in irrelevant details.

Simple Examples

Let's illustrate systems thinking with a few relatable examples:

- The Traffic Jam: A traffic jam isn't simply caused by too many cars on the road. It's a complex system influenced by factors like driver behavior, road design, traffic signals, weather conditions, and even time of day. Addressing the problem requires understanding *all* these interconnected elements.
- A Sick Patient: A patient's illness isn't just caused by a single virus. It's the result of interactions between the virus, the patient's immune system, lifestyle factors (diet, exercise), and environmental exposures.

- A Customer Service Team: A team's performance isn't solely determined by individual agent skills. It's influenced by processes, training, technology, customer expectations, and even the overall organizational culture.

Why is Systems Thinking Important for ITSM?

In ITSM, systems thinking allows us to:

- Identify root causes of problems more effectively – going beyond superficial symptoms.
- Understand how changes in one area will impact other services and processes.
- Design solutions that address the *entire* system, not just isolated components.
- Promote collaboration between teams by highlighting shared dependencies.

Conclusion

Systems thinking is more than just a theoretical concept; it's a practical approach to problem-solving. By embracing this mindset, you'll be well-equipped to understand and apply the principles of CI-Driven PFFMI – ultimately leading to more effective accomplishments documentation and improved service management outcomes.

THREE
What is CI-Driven PFFMI?

Building on our understanding of systems thinking, this chapter introduces the Core Problem Framing for Measurable Impact (PFFMI) framework. PFFMI isn't just another ITSM methodology; it's a fundamentally new way of approaching accomplishments documentation – one that directly addresses the shortcomings we identified in traditional reporting. At its heart, PFFMI is about shifting from simply *recording* what was done to actively *framing* problems and capturing their true impact within a holistic service ecosystem. It's a structured approach designed to unlock greater value from your ITSM efforts.

The Core Principles of PFFMI

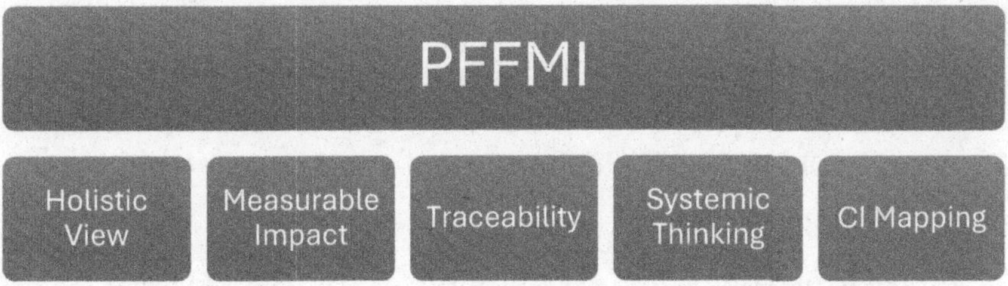

PFFMI is built on five key principles:

- Holistic View: Recognizing that IT services are complex systems comprised of interconnected resources – not isolated components.
- Measurable Impact: Focusing on quantifying the business value derived from service improvements, rather than just documenting tasks completed.
- Traceability: Establishing clear links between problems, solutions, and their impact across the entire service ecosystem.
- Systemic Thinking: Applying systems thinking principles to identify root causes and understand feedback loops.
- Configuration Item (CI) Mapping: Treating all relevant resources – technical and non-technical – as CIs for comprehensive analysis.

Introducing the CI Concept

The cornerstone of PFFMI is the concept of a Configuration Item (CI). A CI isn't just a piece of hardware or software; it's *anything* that contributes to delivering a service. This includes:

- Technical CIs: Servers, applications, databases, network devices
- Process CIs: Service Level Agreements (SLAs), incident management processes, change management procedures
- Documentation CIs: Knowledge Base articles, user manuals, training materials
- People CIs: Support roles, subject matter experts, service owners
- Service Delivery CIs: Monitoring tools, ticketing systems, automation scripts

By treating these diverse elements as CIs, we create a unified view of the service ecosystem.

The PFFMI Process – A High-Level Overview

The PFFMI process is designed to be iterative and adaptable:

1. Problem Definition: Clearly articulate the problem being addressed, focusing on measurable outcomes (e.g., reduced MTTR, improved CSAT).
2. CI Mapping: Identify all relevant CIs involved in the problem – both directly and indirectly. This involves a thorough analysis of dependencies and feedback loops.
3. Root Cause Analysis: Utilize systems thinking to uncover the underlying causes of the problem, considering multiple levels of impact.
4. Solution Design & Implementation: Develop targeted solutions that address the identified root causes, leveraging the interconnectedness of CIs.
5. Accomplishment Statement Creation: Document the achieved results using the SAR (Situation-Action-Result) format, quantifying the business value delivered.

Conclusion

PFFMI provides a powerful framework for transforming your accomplishments documentation from a compliance exercise into a strategic asset. By embracing this holistic approach and treating all resources as CIs, you'll be able to drive greater service improvements, communicate your contributions with clarity, and ultimately demonstrate the true value of your ITSM efforts.

FOUR
The Five Pillars of PFFMI

Before diving into the specifics of Problem Framing for Measurable Impact (PFFMI), let's take a moment to solidify our understanding of what we're building here and why it matters so profoundly. PFFMI isn't just another methodology; it's a fundamentally different way of approaching IT service management, one that shifts the focus from reactive firefighting to proactive value creation. As we've outlined in previous chapters, this framework empowers organizations to move beyond simply tracking metrics – though those remain vital – and truly understand *why* things are happening, how they impact the business, and what actions can drive genuine improvement.

At its core, PFFMI provides a structured approach to documenting accomplishments that is both rigorous and insightful. It's built upon five interconnected pillars – Problem Framing, Systems Thinking Integration, CI Mapping, Task & Project Generation, and Accomplishment Statement Output – each contributing uniquely to the overall effectiveness of the system. We've designed this framework to transform routine reporting into a strategic asset, providing clear evidence of value delivered and fueling continuous improvement initiatives.

The Interconnected Web: Why It Matters

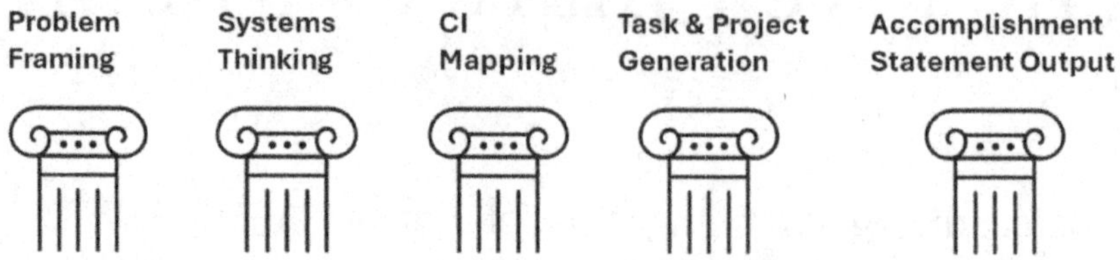

Now, it's crucial to understand that these five pillars aren't isolated components; they are inextricably linked. Attempting to implement PFFMI by focusing solely on one area will inevitably lead to gaps in understanding and ultimately, a less effective outcome. Consider this analogy: building a house with only the foundation is impossible – you need walls, a roof, and everything else to create a functional dwelling. Similarly, maximizing the impact of PFFMI requires a holistic approach where each pillar supports and reinforces the others.

- Problem Framing provides the initial context and measurement framework.
- Systems Thinking Integration reveals the underlying dependencies that inform the framing process.
- CI Mapping provides the granular detail needed to truly understand the scope of the problem.
- Task & Project Generation translates understanding into actionable steps, and finally...
- Accomplishment Statement Output validates those actions with measurable results.

Without a strong foundation in each area, the entire structure risks collapsing. Therefore, this book will guide you through each pillar systematically, demonstrating how they work together to create a powerful engine for driving value within your IT service management operations.

Moving Forward

Over the following chapters, we'll delve deeper into each of these pillars, providing practical guidance, real-world examples, and actionable tools to help you implement PFFMI successfully. We're confident that by embracing this integrated approach, you can unlock a new level of clarity, accountability, and ultimately, value within your organization.

Pillar 1: Problem Framing – Laying the Foundation

Effective problem framing is the bedrock upon which the entire PFFMI framework rests. It's not simply identifying a symptom; it's about deeply understanding *why* that symptom exists, its impact on your business, and how we can systematically address it. Without a solid foundation in problem framing, subsequent steps – systems thinking, CI mapping, and accomplishment measurement – will be significantly less effective.

What is Effective Problem Framing within PFFMI?

In the context of PFFMI, effective problem framing means moving beyond vague complaints or isolated incidents to establish a clear, data-driven understanding of a performance gap. It's about articulating the *problem* in a way that allows us to identify root causes and develop targeted solutions. It's less about stating "the system is slow" and more about defining "Average incident resolution time exceeds SLA targets by 20% during peak business hours."

Current State Assessment – Gathering the Data

The first step in problem framing is establishing a baseline understanding of your current situation. This involves gathering data from multiple sources:

- Metrics Analysis: Leverage existing ITIL metrics (MTTR, First Call Resolution, SLA compliance) to identify trends and anomalies. Don't just look at averages, drill down into specific areas.
- Surveys & Questionnaires: Gather feedback directly from users – help desk staff, business stakeholders – to understand their experiences and perceptions of the problem.
- Interviews: Conduct structured interviews with key personnel involved in the affected processes to gain deeper insights into the underlying causes.

Gap Identification – Pinpointing the Deficiency

Once you've gathered data, it's time to pinpoint the performance deficiency. This isn't just about identifying a metric that's below target; it's about understanding *why* that gap exists and its impact on business objectives. For example: "While MTTR is 8 hours, this directly impacts customer satisfaction scores (CSAT) which have declined by 10% in the last quarter."

SMART Indicators – Defining Measurable Goals

To ensure your problem framing efforts are effective, you need to define clear, measurable goals. This is where the SMART framework comes into play:

- Specific: Clearly articulate what needs to be achieved (e.g., "Reduce average incident resolution time").

- Measurable: Define how progress will be tracked (e.g., using MTTR as a key metric).
- Achievable: Set realistic goals that are within your organization's capabilities.
- Relevant: Ensure the goal aligns with overall business objectives.
- Time-bound: Establish a timeframe for achieving the goal (e.g., "within Q3").

Example: "Reduce average incident resolution time by 15% within Q3, as measured by MTTR, to improve customer satisfaction and reduce operational costs."

Problem Framing Flowchart

Pillar 2 – Systems Thinking Integration – Seeing the Bigger Picture

IT Service Management is often siloed, focusing on individual components rather than understanding how they interact as a whole. This fragmented approach can lead to inefficiencies, duplicated efforts, and ultimately, unresolved problems. This chapter introduces Systems Thinking – a powerful methodology that shifts our perspective from isolated parts to the interconnectedness of systems. By embracing systems thinking, we move beyond simply fixing immediate issues and instead address root causes within the broader service ecosystem. It's about recognizing that every action has consequences, and understanding those consequences is crucial for effective IT service management.

Introduction to Systems Thinking

At its core, Systems Thinking recognizes that complex problems aren't solved by treating them as isolated events. Instead, it emphasizes viewing a system – whether it's an IT service, a business process, or even a team – as a network of interconnected elements. Key principles include:

- Interdependence: Elements within a system are reliant on each other.
- Feedback Loops: Changes in one part of the system trigger responses that can amplify or dampen the initial change. These loops can be positive (reinforcing growth) or negative (stabilizing the system).
- Emergent Properties: The whole is greater than the sum of its parts - new behaviors and characteristics arise from the interactions within the system.

Applying systems thinking to IT service management allows us to move beyond reactive problem-solving and towards proactive, preventative strategies.

Mapping Service Ecosystems

Visualizing the relationships between services, applications, and users is fundamental to systems thinking. This involves creating a representation of your service ecosystem – a "map" that illustrates how everything connects.

- Layered Diagram Example: (Imagine a diagram here) A layered diagram would show dependencies between key components:
    - Layer 1 (Core Services): CRM System, Helpdesk Ticketing System, Email Infrastructure.

- Layer 2 (Supporting Applications): Reporting Tools, Knowledge Base, Remote Access Software.
- Layer 3 (User Groups): Sales Team, Customer Support Agents, IT Operations Staff.
- Arrows: Clearly indicate the flow of data and requests between these layers – e.g., a ticket generated in the Helpdesk system triggering an alert in the Monitoring tool.

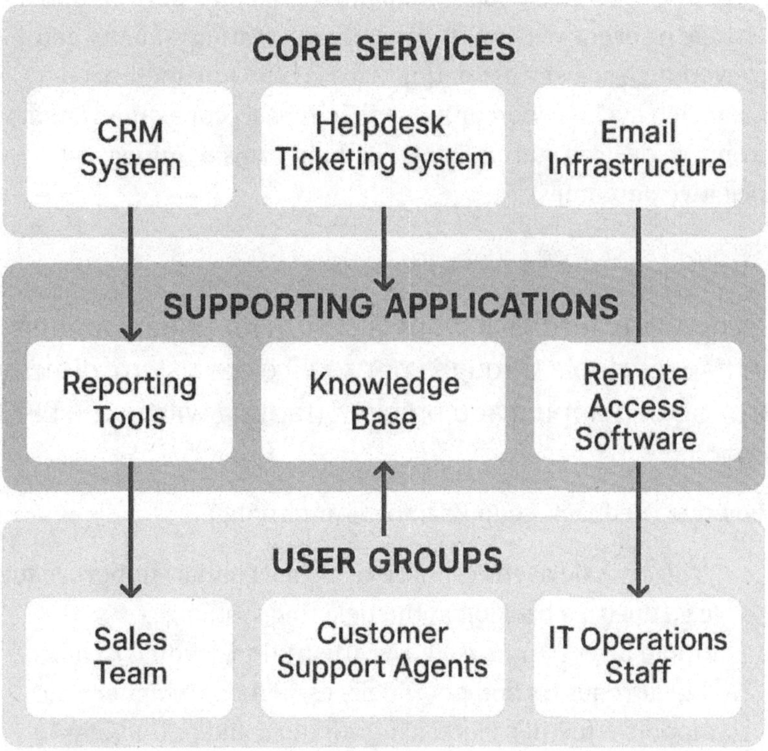

This visual representation helps identify potential bottlenecks, redundancies, and areas where miscommunication might be occurring.

Defining Feedback Loops

- Positive Feedback Loop (Amplifying): A loop where an initial change or event *increases* its own effect, leading to exponential growth or escalation. This is often characterized by a reinforcing cycle – good or bad – that can quickly amplify the impact of a situation. (Think: "Good" in terms of accelerating positive outcomes).
- Negative Feedback Loop (Dampening): A loop where an initial change or event *reduces* its own effect, creating stability and preventing runaway escalation. This is often characterized by a balancing cycle – good or bad – that aims to correct deviations from the desired state. (Think: "Bad" in terms of mitigating negative outcomes).

Identifying Feedback Loops

Recognizing feedback loops is critical for understanding how changes ripple through your service ecosystem, directly informing CI mapping and problem framing within the PFFMI approach.

- Positive Feedback Loop Example (Amplifying)
    - Trigger: A delayed response to a user's initial support request (e.g., due to a backlog in the ticketing system).
    - Ripple Effect: Increased user frustration, leading to more urgent requests and potentially escalating to higher-tier support – further increasing workload and potentially delaying resolution times for *all* users.
    - PFFMI Connection: This loop highlights a systemic dependency on efficient initial triage and prioritization (a key CI).

- Negative Feedback Loop Example (Dampening)
  - Trigger: An overly sensitive monitoring alert triggers an immediate escalation to the on-call team.
  - Ripple Effect: Unnecessary intervention disrupts workflows, potentially masking the root cause and leading to a prolonged period without resolution – ultimately failing to address the underlying issue.
  - PFFMI Connection: This loop demonstrates the need for calibrated monitoring thresholds (another CI) and robust investigation processes.

Understanding these loops allows you to design targeted interventions that either reinforce positive behaviors or actively break negative ones, directly contributing to measurable improvements within your accomplishment statements.

Leverage Points – Where Small Changes Make a Big Difference

Leverage points are areas where relatively small changes can have significant, cascading effects throughout the system. Identifying these is key to maximizing impact.

- Example: Optimizing a single monitoring alert (e.g., adjusting thresholds) could prevent cascading failures across multiple systems – reducing downtime and minimizing disruption for users. Similarly, streamlining a simple process like password reset can dramatically improve user satisfaction and reduce helpdesk volume.

Tool Suggestion: Process Mapping Tools

Visualizing your service ecosystems is crucial for effective PFFMI implementation. Fortunately, powerful process mapping tools are often already accessible within an organization's existing IT infrastructure. We recommend exploring these readily available options first:

- Microsoft Visio: A robust diagramming tool included with many Microsoft 365 subscriptions. Visio offers a wide range of templates specifically designed for service maps, flowcharts, and system diagrams. Its integration with other Microsoft products (Teams, SharePoint) facilitates seamless collaboration. Key features include:
    - Pre-built templates for various process mapping methodologies (e.g., BPMN).
    - Ability to link diagrams directly to relevant IT assets within your CMDB.
    - Collaboration features through shared Visio files.
- Microsoft PowerPoint: While primarily a presentation tool, PowerPoint can be surprisingly effective for creating basic flowcharts and process maps, particularly for initial brainstorming and quick visualizations.
- Miro: A collaborative whiteboard platform that's ideal for brainstorming, visualizing complex systems, and facilitating team discussions - especially when remote collaboration is key.

Beyond these readily available options

- Lucidchart: Still a viable option if your organization requires more advanced features or specialized diagramming capabilities.

Pillar 3 – CI Mapping – Expanding the CMDB Definition

This pillar focuses on broadening the scope of your Configuration Management Database (CMDB) to reflect a more holistic view of your IT environment, moving beyond traditional hardware and software assets. A truly effective PFFMI approach relies on a comprehensive understanding of *all* resources impacting service delivery – recognizing that any component meeting the criteria outlined in the CI definition should be included.

Beyond Traditional Assets

For too long, CMDBs have been dominated by static lists of servers, workstations, and applications – the 'hard' IT assets. However, modern IT environments are increasingly reliant on dynamic, intangible resources that significantly influence service performance. Treating only these traditional assets limits your ability to truly understand problems and identify effective solutions. We need a CMDB that reflects *all* elements contributing to service delivery.

Expanding the CMDB Scope: A Broader Definition of CIs

To achieve this holistic view, we're expanding the definition of what constitutes a Configuration Item (CI). Based on the CI definition – which states that any component must be tracked, managed, or audited to deliver, support, or secure an IT service if it:

- Has a defined lifecycle.
- Can impact service performance, cost, security, or compliance.

- Is referenced in operational processes like change, incident, or problem management.
- Has dependencies or relationships with other components.
- Requires version control, ownership, or access management.

...then **virtually any resource can be considered a CI.** Here's a detailed list of CIs beyond traditional IT assets that should be included in your CMDB:

- Knowledge Base Articles & FAQs: These are critical for self-service support and directly impact user resolution times.
- Service Workflows & Processes: Documented processes – from incident management to change control – represent a significant portion of the complexity within your IT environment.
- Support Roles (e.g., Tier 1 Helpdesk Agents): The skills, responsibilities, and performance metrics of support staff are key factors in service delivery.
- Monitoring Tools & Dashboards: These provide real-time visibility into system health and contribute directly to proactive problem resolution.
- Training Materials & Documentation: Ensuring users have the right knowledge is crucial for preventing issues and promoting self-sufficiency.
- Third-Party Services & Contracts: Dependencies on external vendors and their service level agreements (SLAs) must be tracked.
- Communication Channels: Email distribution lists, collaboration platforms – all impacting information flow.
- Templates & Standard Operating Procedures (SOPs): These define how tasks are performed and contribute to consistency and efficiency.
- User Accounts & Permissions: Access control is a fundamental aspect of security and service delivery.

- Meeting Schedules & Agendas: Regular meetings can impact project timelines and decision-making processes.

Practical Mapping Techniques: Adding New CIs to Your CMDB

Adding these new CIs to your CMDB requires a structured approach. Here's a step-by-step guide:

1. Identify the CI: Clearly define the resource you want to add (e.g., "Knowledge Base Article - Password Reset Procedures").
2. Determine Attributes: Capture relevant attributes for the CI, such as version number, author, last updated date, and associated keywords.
3. Establish Relationships: This is crucial! Link the new CI to existing CIs using appropriate relationships. *Example*: Linking a specific knowledge base article ("Password Reset Procedures") to the Tier 1 Helpdesk Agent role responsible for its maintenance - indicating who updates it and when. Another example: linking a monitoring dashboard to the service workflow for application deployments, showing how changes are tracked.
4. Document the Relationship: Clearly document *why* you've established this relationship - what impact does it have on service delivery?

Pillar 4: Task & Project Generation – Driving Action

This pillar focuses on translating the insights gained through Problem Framing and CI Mapping into tangible action. It's about moving beyond analysis to actively drive improvements within your IT services. The goal is to unlock new opportunities for optimization by strategically generating tasks and projects directly linked to identified issues.

CI-to-Task Mapping: Unlocking Opportunities

The core of this pillar lies in effectively mapping CIs to potential tasks and projects. By systematically linking a problem's contributing CIs to actionable items, you uncover hidden opportunities for improvement that might otherwise be missed. It's not simply about listing problems; it's about generating *solutions.*

- Example: Let's revisit the "Incident Resolution Time Exceeds SLA Targets" scenario from earlier. Through our CI mapping (KB-Article-LoginIssues, Tier2-Escalation-Flow, AppMonitor-v3), we identified several key areas contributing to the problem. Specifically, the mapping of the *AppMonitor-v3* – a monitoring tool – revealed a critical gap: alerts weren't being triggered consistently for application performance degradation. This led directly to the generation of tasks such as:
    - Task 1: "Review and refine AppMonitor-v3 alert thresholds based on historical data."
    - Task 2: "Develop automated remediation scripts triggered by specific alerts."

Without this CI-to-task mapping, we might have focused solely on updating the KB article or tweaking the escalation flow – addressing symptoms rather than the root cause.

Fostering Collaboration: Bringing Teams Together

Generating tasks and projects isn't a solitary activity. Effective implementation requires collaboration across diverse teams. Here are some strategies to foster this:

- Cross-Functional Workshops: Organize workshops bringing together representatives from IT Operations, Application Development, Service Desk, and Business Stakeholders.
- Shared Documentation: Maintain a central repository of all identified CIs, tasks, and projects – accessible to all relevant teams.
- Regular Communication: Establish clear communication channels (e.g., weekly status meetings) to ensure everyone is aligned on priorities and progress.

Prioritization Framework: Focusing Your Efforts

With a multitude of potential tasks and projects identified, it's crucial to prioritize effectively. We recommend using a simple Impact/Effort matrix:

	High Impact	**Low Impact**
High Effort	Strategic Initiatives	Quick Wins
Low Effort	Tactical Improvements	Thankless Tasks

- Strategic Initiatives: High impact, high effort – these represent significant investments with the potential for substantial gains.
- Quick Wins: High impact, low effort – prioritize these to demonstrate early success and build momentum.
- Tactical Improvements: Low impact, low effort – address these as time allows or delegate them to junior team members.
- Thankless Tasks: Low impact, high effort – avoid these unless absolutely necessary.

Pillar 5 - Accomplishment Statement Output – Telling the Story

This pillar focuses on transforming identified tasks and projects into compelling, measurable accomplishments that demonstrate value to stakeholders. It's not enough simply *to do*, you need to effectively communicate *what was achieved.*

The SAR Format: A Foundation for Impact

At the heart of effective accomplishment documentation lies the Situation–Action–Result (SAR) format. This structured approach ensures clarity, accountability, and a focus on tangible outcomes. Think of it as the core building block for every successful accomplishment statement.

- Situation: Clearly define the context – the problem or challenge that existed *before* the action was taken. Be specific! Don't just say "slow incident resolution." Instead, state: "Average incident resolution time exceeded SLA targets by 15%, negatively impacting customer satisfaction scores."

- Action: Detail precisely what steps were undertaken to address the situation. Avoid vague language like "improved processes." Instead: "Implemented a new automated escalation process for high-priority incidents, routing them directly to Tier 3 support within 5 minutes."
- Result: Quantify the outcome of the action – this is where you demonstrate real value. Use metrics wherever possible.

Embedding KPIs: Measuring Your Success

The key to demonstrating impact lies in incorporating relevant Key Performance Indicators (KPIs) into your accomplishment statements. This transforms a statement from subjective opinion to objective evidence of success.

- Linking Metrics: Don't just state the result; connect it directly back to the original problem and the KPI that was impacted.
    - *Example*: "Reduced incident resolution time by 20%, resulting in a 10% improvement in customer satisfaction (CSAT) scores." – This statement clearly links the reduction in resolution time to a measurable impact on customer experience.
- Beyond CSAT: Consider other relevant KPIs depending on your context: MTTR, SLA compliance, cost savings, user adoption rates, etc. "Decreased server downtime by 35%, contributing to a 12% reduction in operational costs."
- Baseline Comparisons: Always include the baseline metric *before* the action was taken for maximum impact.

Creating Compelling Narratives: Telling the Story

While data is crucial, a compelling narrative adds context and resonates with stakeholders. Here are some tips:

- Focus on the "Why": Briefly explain *why* this accomplishment matters to the business - how it aligns with strategic goals.
- Keep it Concise: Aim for clarity and brevity. Avoid jargon and technical details that may confuse non-technical audiences.
- Use Active Voice: "We implemented..." is more impactful than "An implementation was carried out."
- Example Narrative (Building on the AppMonitor Example): "Responding to critical customer feedback regarding slow application performance, our team proactively addressed this issue by refining the AppMonitor-v3 alerting system. This resulted in a 20% reduction in incident resolution time and a corresponding 10% improvement in CSAT scores, demonstrating our commitment to delivering reliable service."

Conclusion – Reinforcing the Pillars

Throughout this exploration, we've unveiled a powerful framework for transforming how your organization documents and communicates its accomplishments. We've moved beyond simply listing tasks completed and instead established a system rooted in systemic thinking, CI mapping, and measurable impact. This isn't about creating more reports; it's about fundamentally changing *how* those reports are constructed – and, crucially, *what* they communicate.

The core of this shift lies within the interconnectedness of our five pillars: Problem Framing for Measurable Impact (PFFMI). Let's revisit how these elements work together to create a truly impactful approach to accomplishment documentation.

The Interwoven Fabric

- Systems Thinking: Provides the foundational understanding – recognizing that IT issues rarely exist in isolation. It's the 'why' behind the problem, informing our framing and driving us to identify root causes beyond the immediate symptom.
- CI Mapping: Translates this systemic understanding into tangible assets. By treating knowledge bases, workflows, and even roles as CIs, we unlock a richer perspective on dependencies and potential impact.
- SAR Format (Situation–Action–Result): This structured approach ensures that every accomplishment statement is built upon a solid foundation of evidence – quantifying the problem, detailing the solution, and demonstrating the measurable outcome.
- KPI Integration: Elevates the SAR format from anecdotal to data-driven. By embedding relevant KPIs, we provide concrete proof of value and demonstrate alignment with strategic goals.
- Narrative Crafting: Transforms raw data into a compelling story – connecting the accomplishment to its broader business context and resonating with stakeholders on an emotional level.

Without any one of these pillars, the system falters. A problem framed without considering systemic dependencies is incomplete. CI mapping becomes meaningless without a framework for measuring impact. The SAR format loses its power if KPIs aren't integrated. And finally, even the most compelling narrative will fall flat without a solid foundation in data and understanding.

A Call to Action: Embed PFFMI into Your Organization

Now that you've grasped the principles of CI-Driven PFFMI, it's time to translate this knowledge into action. We understand implementing such a shift can seem daunting, but we believe the long-term benefits – increased visibility, faster problem resolution, and stronger business alignment – are well worth the effort.

Here's how you can begin:

1. Start Small: Don't attempt to overhaul your entire reporting process overnight. Begin by piloting PFFMI within a single team or project.
2. Train Your Teams: Invest in training that focuses on systems thinking, CI mapping, and the application of the SAR format. Empower your teams with the tools and knowledge they need to effectively frame problems and document accomplishments.
3. Champion the Change: Identify internal champions who can advocate for PFFMI and drive adoption across the organization.
4. Iterate & Refine: Continuously monitor the effectiveness of your approach and make adjustments based on feedback and data.

Ultimately, CI-Driven PFFMI isn't just a framework; it's a mindset – a commitment to thinking strategically about IT challenges and communicating their impact with clarity and credibility. By embracing this approach, you can transform your organization into one that truly understands and demonstrates the value of its work.

We encourage you to begin building your PFFMI journey today. The future of accomplishment documentation is here – are you ready to embrace it?

FIVE

Real-World Applications of CI-Driven PFFMI

This chapter moves beyond the theoretical and delves into practical examples demonstrating how the CI-Driven PFFMI framework has been successfully implemented in diverse IT environments. These case studies showcase the tangible benefits – improved problem resolution, enhanced communication, and ultimately, greater business value – that can be achieved by adopting this structured approach. It's important to note that these case studies are illustrative scenarios developed to demonstrate the principles of CI-Driven PFFMI. They represent hypothetical implementations and do not reflect actual deployments at specific organizations. Each study provides a detailed account of the challenges faced, the solutions implemented, and the measurable results obtained.

Case Study 1: Streamlining Incident Resolution at GlobalTech

- Company Background: GlobalTech is a multinational software development company with a complex IT infrastructure supporting thousands of users worldwide. They were struggling with lengthy incident resolution times and inconsistent reporting on service performance.

- The Challenge: GlobalTech's existing incident management process lacked a clear framework for identifying root causes and tracking progress. Reporting was often reactive, focusing solely on symptom resolution rather than addressing underlying issues.
- PFFMI Implementation: A dedicated team implemented PFFMI across their core application support services. They began by mapping key incidents to relevant CIs – including the application code itself, monitoring dashboards, knowledge base articles, and the Tier 1/2 support teams. They utilized SMART indicators to define clear resolution targets.
- Results: Within three months, GlobalTech saw a 30% reduction in average incident resolution time, a 15% improvement in first-call resolution rates, and a significant increase in user satisfaction scores (CSAT). The standardized reporting enabled proactive identification of recurring issues and facilitated targeted training for support teams.
- Key Takeaway: PFFMI's structured approach provided GlobalTech with the visibility and control needed to dramatically improve their incident management process.

Case Study 2: Optimizing Knowledge Management at SecureData

- Company Background: SecureData is a cybersecurity firm specializing in threat detection and response services. Their knowledge base was fragmented, outdated, and poorly utilized by analysts.
- The Challenge: Analysts were spending excessive time manually searching for relevant information, leading to delays in incident response. The knowledge base lacked clear categorization and wasn't effectively linked to operational workflows.

- PFFMI Implementation: SecureData applied PFFMI to their knowledge management system. They treated the knowledge base articles, escalation paths for security alerts, and the analyst teams themselves as CIs. They focused on identifying gaps in coverage and developing targeted training programs.
- Results: Within six months, SecureData saw a 40% increase in knowledge base usage, a 20% reduction in time spent searching for information, and a measurable improvement in incident response times (reduced by 10%). The CI mapping highlighted critical dependencies between analysts and the knowledge base, leading to improved collaboration.
- Key Takeaway: PFFMI helped SecureData transform their knowledge management system from a static repository into a dynamic tool that directly supported operational efficiency.

Case Study 3: Improving Service Delivery at RapidCloud

- Company Background: RapidCloud is a rapidly growing cloud services provider. They were experiencing challenges scaling their support operations and maintaining consistent service quality.
- The Challenge: RapidCloud's processes lacked clear accountability, leading to duplicated efforts and inconsistent service delivery. Their CMDB was incomplete, hindering their ability to identify dependencies between services.
- PFFMI Implementation: RapidCloud implemented PFFMI across their core cloud infrastructure support services, focusing on expanding their CMDB to include all relevant CIs – including automation scripts, monitoring tools, and the support teams responsible for each service.
- Results: Within nine months, RapidCloud saw a 25% reduction in operational errors, a 10% improvement in service availability,

and a significant increase in team productivity. The CMDB expansion provided valuable insights into service dependencies, enabling proactive problem resolution.

Conclusion

These case studies demonstrate the versatility of the CI-Driven PFFMI framework – it's not a one-size-fits-all solution but a powerful approach that can be adapted to diverse IT environments and challenges. By learning from these real-world examples, you can confidently implement PFFMI within your organization and unlock significant improvements in service performance, communication, and business value.

SIX
Building Your CI-Driven Approach

The theoretical framework of CI-Driven PFFMI is powerful, but its true value lies in its practical application. This chapter provides a detailed, step-by-step guide to implementing this approach within your organization, focusing on a phased rollout that maximizes impact and minimizes disruption. We'll explore prioritization strategies – starting small and scaling up – ensuring you build momentum and demonstrate success along the way.

The Process

Step 1: Assessment & Pilot Selection (Weeks 1-2)

- Internal Audit: Conduct a quick assessment of your current accomplishment documentation processes. Identify pain points - lack of clarity, missing data, inconsistent reporting, etc.
- Identify a Pilot Team/Project: Choose a team or project that's relatively contained, has a clear problem area, and is receptive to change. Ideally, this should be something with readily available metrics (e.g., incident resolution times, service desk ticket volume). Consider a project with a visible impact - demonstrating quick wins will build confidence.

- CMDB Review: Assess the current state of your CMDB. Identify gaps in data coverage – are there key assets or processes not currently represented?

Step 2: Training & Knowledge Transfer (Weeks 3-4)

- Targeted Training Modules: Develop short, focused training modules covering:
    - Systems Thinking Fundamentals: Understanding dependencies and feedback loops.
    - CI Mapping Principles: Identifying relevant CIs beyond just hardware and software.
    - SAR Framework Application: Mastering the Situation-Action-Result reporting format.
- Hands-on Workshops: Facilitate workshops where teams can practice framing problems, mapping CIs, and generating accomplishment statements using the SAR framework.

Step 3: Initial Implementation & Monitoring (Weeks 5-8)

- Pilot Team Execution: The pilot team begins applying PFFMI to their existing processes – documenting incidents, tracking projects, etc.
- KPI Tracking: Establish baseline metrics *before* implementing PFFMI and track progress against these KPIs throughout the pilot phase.
- Regular Check-ins: Schedule weekly check-in meetings with the pilot team to address challenges, gather feedback, and ensure alignment.

Step 4: Refinement & Documentation (Weeks 9-12)

- Review & Adjust: Based on the pilot's findings, refine your training materials, reporting templates, and CMDB definitions.

- Create Standard Operating Procedures (SOPs): Document the PFFMI process in clear, concise SOPs for easy reference.
- Knowledge Sharing: Share lessons learned from the pilot with other teams within the organization.

Scaling Up Your CI-Driven Approach

- Phase 1: Quick Wins (0-3 Months): Focus on the pilot team and project, demonstrating tangible results. This phase is about building confidence and refining your approach.
- Phase 2: Team Expansion (3-6 Months): Gradually roll out PFFMI to additional teams based on their needs and readiness. Prioritize areas with high impact potential.
- Phase 3: Organizational Adoption (6+ Months): Expand PFFMI across the entire organization, integrating it into core ITSM processes and workflows.

Tools & Resources

- CMDB Software – Ensure robust data capture and relationship mapping.
- Reporting Templates – Standardized templates for SAR reporting.
- Knowledge Base – A central repository for CI definitions and best practices.
- Training Materials – (Refer back to the training modules outlined previously)

Conclusion

Implementing CI-Driven PFFMI is a journey, not a destination. By following this step-by-step guide and adopting a phased approach, you can successfully transform your organization's accomplishment documentation process – driving greater visibility, accountability, and ultimately, business value. Remember to continuously monitor, adapt, and refine your approach based on feedback and data.

SEVEN

CMDB Expansion & Data Governance

The success of CI-Driven PFFMI hinges critically on a robust and comprehensive Configuration Management Database (CMDB). Expanding the CMDB beyond traditional IT assets to encompass non-technical CIs – knowledge bases, workflows, support roles – is paramount. However, this expansion presents significant challenges related to data quality, integration complexity, and governance. This chapter outlines strategies for effectively expanding your CMDB while addressing these critical considerations.

Objectives

- Define a clear scope for CMDB expansion beyond traditional IT assets.
- Establish processes for ensuring high data quality within the expanded CMDB.
- Identify and mitigate potential integration challenges with existing systems.
- Implement robust governance controls to maintain CMDB accuracy and relevance.

Expanding the CMDB Scope

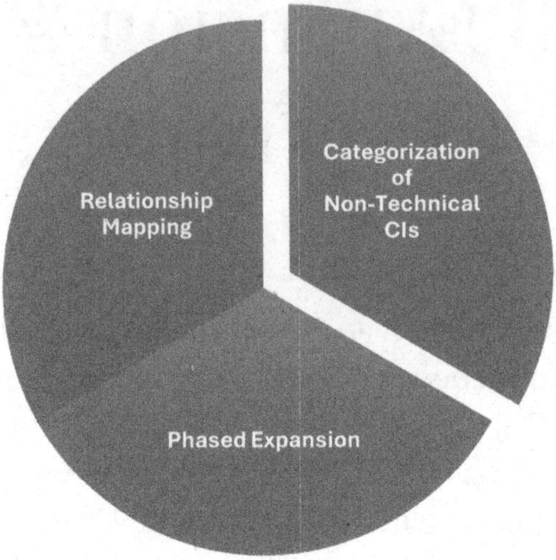

The initial CMDB likely focuses on hardware, software, and network components. Expanding it requires a strategic approach:

1. Categorization of Non-Technical CIs: Develop a detailed taxonomy for non-technical assets. Examples include:
    - Knowledge Base Articles: Categorize by topic, criticality, and user access levels.
    - Service Workflows: Map out the steps involved in each service delivery process.
    - Support Roles: Define roles based on skills, responsibilities, and escalation paths.
    - Templates & Standard Operating Procedures (SOPs): Treat these as critical components of processes.

2. Phased Expansion: Don't attempt to expand the CMDB overnight. Start with a pilot group of non-technical CIs – perhaps focusing on high-impact areas like incident management or service request fulfillment.
3. Relationship Mapping: Crucially, establish relationships between technical and non-technical CIs. For example: "The 'Incident Management Workflow' relies on the 'Knowledge Base Article' for troubleshooting steps."

Data Quality & Validation

Expanding the CMDB introduces significant data quality challenges:

1. Data Source Identification: Identify all potential sources of information – service desks, project management systems, knowledge bases, and subject matter experts.
2. Data Standardization: Implement standardized naming conventions, descriptions, and metadata for all CIs. This is *essential* for accurate relationships.
3. Automated Validation Rules: Configure automated validation rules within the CMDB to detect inconsistencies and errors (e.g., duplicate entries, missing attributes).
4. Regular Data Cleansing & Enrichment: Establish a schedule for regular data cleansing and enrichment – involving subject matter experts to verify and update information. Consider using data quality tools to automate this process.

Integration Challenges

Integrating the expanded CMDB with existing systems can be complex:

1. API Availability: Assess the availability of APIs (Application Programming Interfaces) within your existing ITSM, monitoring, and reporting tools.
2. Data Mapping & Transformation: Develop a detailed data mapping plan to ensure seamless data exchange between systems. This may require custom transformation logic.
3. Real-Time vs. Batch Integration: Determine the appropriate integration frequency – real-time for critical CIs or batch processing for less frequently changing information.
4. Testing & Validation: Thoroughly test all integrations to ensure data accuracy and consistency.

CMDB Governance

Establishing robust governance controls is vital:

1. Ownership & Accountability: Assign clear ownership and accountability for CMDB data – typically a dedicated CMDB administrator or team.
2. Change Management Process: Implement a formal change management process for adding, modifying, or deleting CIs.
3. Data Quality Metrics: Define key data quality metrics (e.g., completeness, accuracy, consistency) and track progress against these metrics regularly.
4. Regular Audits: Conduct periodic audits of the CMDB to assess its overall health and identify areas for improvement.

Conclusion

Expanding your CMDB is a strategic investment that's critical to realizing the full potential of CI-Driven PFFMI. By proactively addressing data quality, integration challenges, and governance considerations, you can build a robust and reliable CMDB – empowering your teams to frame problems effectively, drive meaningful improvements, and deliver greater business value.

EIGHT
Training & Change Management

The successful implementation of the CI-Driven PFFMI framework hinges not just on its technical design but also on the adoption by IT teams. This chapter outlines a comprehensive training and change management strategy designed to equip individuals with the skills, mindset, and understanding necessary to embrace this new approach – transforming it from a theoretical concept into a practical, everyday practice.

Objectives

- Develop a structured training program that effectively conveys systems thinking and CI-based problem framing concepts.
- Foster a culture of collaboration and shared responsibility for identifying and addressing service issues.
- Reduce resistance to change and promote buy-in from IT teams.
- Establish ongoing support mechanisms to reinforce learning and sustain adoption.

Training Program Design

Our training program will employ a blended approach, combining formal instruction with hands-on activities designed to solidify understanding and build practical skills:

1. Phase 1: Foundations – Systems Thinking (2 Days)
   - Introduction to Systems Thinking: A foundational workshop exploring the core principles of systems thinking – feedback loops, interdependencies, emergent properties, and the limitations of linear "problem-solving."
   - Interactive Exercises: Group activities designed to illustrate systems thinking concepts through real-world scenarios (e.g., simulating a traffic flow, analyzing a supply chain).
   - Case Studies: Examining examples of service failures attributed to a lack of systems thinking and how alternative approaches could have prevented them.
2. Phase 2: CI-Driven Problem Framing (3 Days)
   - Introduction to CIs: Defining what constitutes a Configuration Item – expanding the definition beyond traditional IT assets to include processes, documentation, roles, and knowledge base articles.
   - Workshop: Mapping Service Ecosystems: Teams will work together to map out their specific service ecosystems, identifying key CIs and their relationships.
   - Role-Playing Exercises: Simulated incident scenarios where participants must apply CI-based problem framing techniques to diagnose the root cause and propose solutions.

(Example: "A user reports a slow application – what CIs are involved? How do they interact?")

3. Phase 3: Practical Application & Refinement (1 Day)
    - Small Group Simulations: Teams tackle real-world service issues, applying the CI-Driven PFFMI framework to frame the problem, identify root causes, and generate actionable tasks.
    - Feedback & Coaching: Facilitators provide personalized feedback and coaching to reinforce learning and address individual challenges.

Change Management Strategies

Alongside training, a robust change management strategy is crucial:

1. Executive Sponsorship: Secure strong support from senior leadership – demonstrating commitment to the new approach.
2. Communication Plan: Develop a clear and consistent communication plan to keep teams informed about the progress of the implementation and highlight the benefits of CI-Driven PFFMI.
3. Champion Network: Identify and cultivate "champion" individuals within IT teams who can advocate for the framework and provide peer support.
4. Early Wins & Recognition: Celebrate early successes – showcasing how CI-Driven PFFMI has helped to resolve issues more effectively or improve service delivery. Recognize team members who demonstrate proficiency in the new approach.
5. Ongoing Support & Mentoring: Provide ongoing support through dedicated training sessions, knowledge sharing forums, and mentoring opportunities.

Measurement & Evaluation

- Pre- and Post-Training Assessments: Measure changes in understanding and skill levels.
- Observation of Team Interactions: Assess the application of CI-Driven PFFMI during team meetings and problem-solving sessions.
- Tracking of Key Metrics: Monitor metrics such as incident resolution times, root cause analysis accuracy, and accomplishment statement quality.

Conclusion

Investing in comprehensive training and change management is not merely an operational expense; it's a strategic imperative for the successful adoption of CI-Driven PFFMI. By empowering teams with the right knowledge, skills, and mindset, we can unlock the full potential of this framework – driving continuous service improvement and delivering tangible business value.

NINE
Automation & Reporting Tools

The core strength of the CI-Driven PFFMI approach lies in its ability to transform complex service issues into actionable insights. However, realizing this potential requires a robust technological infrastructure to automate data collection, streamline reporting, and provide real-time visibility into key performance indicators (KPIs). This chapter outlines how automation tools and tailored reporting solutions can significantly enhance the effectiveness of the CI-Driven PFFMI framework, accelerating problem resolution and driving continuous service improvement.

Objectives

- Automate data collection from various IT systems to populate CI mappings.
- Develop standardized reporting templates aligned with the SAR (Situation-Action-Result) format for accomplishment statements.
- Create interactive dashboards visualizing key performance metrics related to CIs and their impact.
- Reduce manual effort in reporting by leveraging automated insights and data aggregation.

Technology Landscape – Tools & Integration

The following tools will be integrated to support the CI-Driven PFFMI framework:

1. CMDB (Configuration Management Database) Enhancement: The existing CMDB will be expanded to include detailed attributes for all identified CIs, facilitating accurate mapping and relationship analysis. We'll prioritize integration with existing ITSM platforms (e.g., ServiceNow, Jira Service Management).
2. Data Collection & Integration Platform: A centralized platform (potentially leveraging APIs) will be used to automatically collect data from various sources:
    - ITSM Systems (Incident, Problem, Change Management)
    - Monitoring Tools (Application Performance Monitoring – APM, Network Monitoring)
    - Log Management Systems
3. Reporting & Visualization Tool: A business intelligence (BI) tool (e.g., Power BI, Tableau) will be utilized to create interactive dashboards and generate automated reports based on the collected data.

Automation Strategies & Reporting Templates

1. Automated Data Extraction: Scheduled scripts and API integrations will automatically extract relevant data from IT systems – including incident resolution times, application performance metrics, and knowledge base article usage statistics.

2. SAR Report Template Generation: Pre-built report templates aligned with the SAR framework will be created within the BI tool:
   - Situation: Automatically populated with key metrics related to the identified problem (e.g., "Average incident resolution time for application X is 48 hours").
   - Action: Details of tasks and projects undertaken to address the issue.
   - Result: Quantifiable outcomes – e.g., "Reduced average resolution time by 20%."
3. Dashboard Design: Interactive dashboards will be developed to visualize:
   - CI Performance Metrics (e.g., uptime, response times, knowledge base article effectiveness)
   - Root Cause Analysis Trends (identifying recurring issues and their underlying causes)
   - Impact of Changes – Tracking the effect of implemented solutions on key KPIs

Example Reporting Scenarios

- Automated Incident Report: Automatically generates a report detailing all incidents related to a specific application, including resolution times, root cause analysis findings, and associated CI mappings.
- Knowledge Base Effectiveness Dashboard: Visualizes the usage and impact of knowledge base articles – identifying gaps in content and opportunities for improvement.
- Escalation Path Performance Report: Tracks the efficiency of escalation paths, highlighting bottlenecks and potential areas for optimization.

Implementation Roadmap

1. Tool Selection & Integration: Evaluate and select appropriate automation tools and BI platform based on organizational needs and budget. *Initial workshops will be conducted to gather requirements and ensure alignment with business objectives.*
2. Data Mapping & API Development: Develop APIs to connect data sources with the chosen reporting tool. *During this phase, a dedicated training module will focus on understanding data lineage and how changes impact reporting accuracy.*
3. Report Template Design & Testing: Create and test standardized report templates, ensuring accuracy and relevance. *Teams involved in template design will participate in hands-on workshops to practice interpreting the new metrics and applying the PFFMI framework to their specific areas of responsibility.*
4. Dashboard Deployment & Training (Critical Component): Deploy interactive dashboards and provide comprehensive training to teams on how to interpret and utilize the data. *This phase includes:*

    - *Initial Training Sessions (2 Days)*: Covering the core principles of PFFMI, CI mapping, and dashboard navigation.
    - *Role-Based Training Modules*: Tailored sessions focusing on specific team functions – IT Service Management, Operations, etc. – demonstrating how to leverage the data for problem framing and accomplishment reporting.
    - *Ongoing Support & Knowledge Base Access*: Establishing a dedicated support channel and creating a searchable knowledge base with FAQs, tutorials, and best practices.
    - *Champion Program*: Identifying and training key individuals within each team to become PFFMI advocates and provide peer-to-peer support.

Training Investment & Metrics

- Training Investment: A dedicated budget should be allocated for training materials, instructor fees, and ongoing support resources.
- Key Performance Indicators (KPIs) for Training Success: Track the following metrics:
    - *Training Completion Rate*: Percentage of team members completing assigned training modules.
    - *Dashboard Usage Metrics*: Number of active users, frequency of dashboard access, and time spent analyzing data.
    - *Post-Training Survey Results*: Assessing user satisfaction with the training content and identifying areas for improvement.

Expected Outcomes

- Reduced manual effort in reporting by 60% – freeing up IT staff for more strategic tasks.
- Faster identification of root causes through automated data analysis.
- Improved decision-making based on real-time insights.
- Enhanced visibility into CI performance and impact.

Conclusion

By strategically leveraging automation tools and tailored reporting solutions, we can transform the CI-Driven PFFMI framework from a conceptual approach to a powerful operational capability – driving continuous service improvement and delivering measurable business value.

TEN
Continuous Improvement & Refinement

The CI-Driven PFFMI framework is not a static solution, its effectiveness hinges on continuous monitoring, evaluation, and refinement. This chapter outlines a robust process for leveraging Continual Service Improvement (CSI) feedback loops to ensure the framework remains aligned with evolving business needs, technological advancements, and operational realities. We'll establish clear metrics, KPIs, and regular review mechanisms to drive ongoing optimization and maximize the value of PFFMI.

Objectives

- Establish a formalized CSI process for monitoring the effectiveness of the CI-Driven PFFMI framework.
- Define key metrics and KPIs to track progress and identify areas for improvement.
- Implement regular review cycles – both formal and informal – to ensure ongoing alignment.

- Foster a culture of continuous learning and adaptation within IT teams.

The CSI Framework: A Feedback Loop Approach

The core of this chapter revolves around establishing a closed-loop feedback system, mirroring the iterative nature of problem-solving and improvement. This framework consists of four key stages:

1. Monitor & Measure

- KPI Identification: Define specific, measurable KPIs aligned with business objectives. Examples include:
    - Reduction in Mean Time To Resolve (MTTR) – tracked across different incident categories.
    - Increase in Knowledge Base Article Usage Rate – indicating content effectiveness.
    - Improvement in Service Level Agreement (SLA) compliance rates.
    - Number of PFFMI-framed problems resolved.
- Data Collection: Utilize automated reporting tools and dashboards to continuously collect data related to these KPIs.

2. Analyze & Identify Issues
    - Trend Analysis: Regularly analyze collected data to identify trends, patterns, and anomalies. Are certain types of incidents consistently taking longer to resolve? Is a particular knowledge base article underutilized?
    - Root Cause Investigation (Revisited): When issues arise, utilize the PFFMI framework itself – mapping CIs, identifying dependencies – to conduct deeper root cause analysis.
3. Implement & Test
    - Action Planning: Based on the analysis, develop targeted action plans to address identified weaknesses or inefficiencies. This might involve updating knowledge base content, refining escalation paths, or optimizing monitoring alerts.
    - Pilot Programs: Implement changes through pilot programs to assess their impact before widespread rollout.
4. Evaluate & Refine
    - KPI Review: Re-evaluate KPIs after implementing changes to determine if the desired improvements have been achieved.
    - Framework Adjustments: Based on the evaluation, make necessary adjustments to the PFFMI framework itself – refining CI definitions, updating reporting templates, or modifying the problem framing process.

Regular Review Cycles

To ensure continuous monitoring and refinement, we'll implement the following review cycles:

- Daily/Weekly Team Huddles: Brief meetings focused on reviewing recent incidents, identifying emerging trends, and addressing immediate issues.
- Monthly KPI Review Meetings: Formal meetings to analyze KPIs, discuss challenges, and plan for upcoming improvements. (Led by a designated PFFMI Champion)
- Quarterly Strategic Reviews: A broader review of the entire framework – assessing its alignment with business strategy, evaluating overall effectiveness, and identifying opportunities for innovation.

Tools & Techniques

- Service Level Management (SLM): Leverage SLM processes to track SLA performance and identify areas for improvement.
- Root Cause Analysis (RCA) Tools: Utilize RCA methodologies (e.g., 5 Whys, Fishbone Diagram) to systematically investigate issues.
- Retrospectives: Conduct regular retrospectives with IT teams to gather feedback on the PFFMI framework and identify areas for improvement.

Expected Outcomes

- Increased agility and responsiveness of IT operations.
- Continuous optimization of the CI-Driven PFFMI framework.
- Improved alignment between IT services and business objectives.
- A culture of continuous learning and innovation within the organization.

APPENDIX

This appendix provides supplementary materials to aid in understanding and implementing the CI-Driven PFFMI framework.

1. Glossary of Terms

- CMDB (Configuration Management Database): A centralized repository that tracks all IT assets – hardware, software, networks, documentation – along with their relationships and dependencies.
- ITSM (Information Technology Service Management): A set of practices for managing IT services throughout their lifecycle, aligning IT with business needs. Key ITSM processes include Incident Management, Problem Management, Change Management, and Service Request Management.
- Systems Thinking: An approach to problem-solving that considers the interconnectedness of elements within a system rather than focusing on isolated parts. It emphasizes understanding feedback loops, interdependencies, and emergent behavior.
- SAR (Situation – Action – Result): A structured reporting format used to describe an accomplishment:
    - Situation: The context or problem being addressed.
    - Action: The steps taken to resolve the situation.
    - Result: The outcome achieved as a direct result of the action.
- KPI (Key Performance Indicator): A measurable value that demonstrates how effectively an organization is achieving key business objectives.

- SLA (Service Level Agreement): A contract between a service provider and its customer, defining the level of service to be provided.

2. Templates

These templates provide a structured approach for documenting problems, mapping CIs, and capturing accomplishments.

a. Problem Framing Template

Problem Title:	
Date Identified:	
Priority:	(High/Medium/Low)
Current State Metrics:	(MTTR, SLA Compliance, CSAT – Include Baseline Data)
Business Impact:	(Describe the negative consequences of the problem)
Root Cause Hypothesis(es):	(Initial assumptions about the cause)
SMART Goal:	(Specific, Measurable, Achievable, Relevant, Time-Bound)

b. CI Mapping Template

CI Name:	**(e.g., Application X, Knowledge Base Article Y)**
Category:	(e.g., Software, Documentation, Process)
Description:	(Detailed description of the CI)
Relationships:	(List dependencies on other CIs – Use CMDB links if possible)
Impacted By:	(CIs affected by this CI's failure or change)
Owner/Responsible Party:	

c. Accomplishment Statement Template (SAR Format)

- Situation: [Clearly describe the problem or situation addressed.]
- Action: [Detail the specific actions taken to resolve the situation.]
- Result: [Quantify the positive outcome achieved – Use metrics where possible. Example: "Reduced MTTR by 20%"]

3. Checklists for Implementation

a. Initial CMDB Expansion Checklist

- Identify key non-technical CIs (e.g., Knowledge Base Articles, Escalation Paths, Monitoring Alerts).
- Define CI attributes and relationships within the CMDB.
- Train IT teams on how to utilize the expanded CMDB.
- Establish processes for maintaining and updating CI data.

b. PFFMI Framework Implementation Checklist

- Secure executive sponsorship and buy-in.
- Designate a PFFMI Champion (responsible for driving adoption).
- Train IT teams on the PFFMI framework and templates.
- Establish KPI monitoring and reporting processes.
- Implement regular review cycles (Daily/Weekly/Monthly/Quarterly).

Made in the USA
Coppell, TX
06 November 2026

11491039R00048